BEEF

FARM TO MARKET

Jason Cooper

Rourke Publications, Inc.
Vero Beach, Florida 32964

Edited by Pamela J.P. Schroeder

PHOTO CREDITS
All photos © Lynn M. Stone

Library of Congress Cataloging-in-Publication Data
Cooper, Jason, 1942-
    Beef / Jason Cooper.
        p.   cm. — (Farm to market)
    Summary: Describes where and how North American beef
cattle are raised, slaughtered, processed, and marketed.
    ISBN 0-86625-617-2
    1. Beef cattle—Juvenile literature. 2. Beef—Juvenile literature.
3. Beef cattle—North America—Juvenile literature. 4. Beef—North
America—Juvenile literature. [1. Beef cattle. 2. Cattle. 3. Beef.]
I. Title. II. Series: Cooper, Jason, 1942- Farm to market.
SF207.C66   1997
841.3'62—dc21   97-13038
                                                      CIP
                                                      AC

Printed in the USA

BEEF

You have probably eaten beef all your life, perhaps without knowing what you were eating!

Beef is the meat of adult cattle. It's served in several ways. The hamburgers you eat, as well as the steaks, are beef.

Beef is a red meat with lines of white fat called marbling. Beef turns gray as it is cooked.

Americans like beef. The average person eats about 70 pounds (32 kilograms) of beef each year.

T-bone steak is a popular cut of beef. The United States is the top beef-producing nation, but people in Argentina eat more beef per person.

BEEF CATTLE

Cattle specially raised for their meat are beef cattle. They have more blocklike bodies than dairy, or milk, cattle. Any kind of cattle, however, can be made into beef.

All cattle are closely related, but farmers have developed several groups of cattle called **breeds** (BREEDZ). Each breed is somewhat different from the others.

Popular beef breeds in North America include Angus, Hereford, Charolais, Brahman, and others.

Many farmers raise mixed-breed cattle, or cattle with parents of two breeds. Mixed breeds often have the best of both pure breeds.

This purebred Hereford cow shows the blocky body beef breeds have.

WHERE BEEF CATTLE ARE RAISED

Beef cattle are raised all through the United States and Canada. Most are in the West and Midwest. Texas, with more than 13 million beef cattle, is the leading beef state. Kansas, Nebraska, Oklahoma, and Missouri follow Texas. In Canada, most beef cattle are raised in Alberta.

Different breeds are raised in different climates. The Brahman and Brahman mixed breeds do well, for example, in hot-weather states, like Florida.

Mixed-breed beef cattle huddle on a wintry day in western Montana. Mixed-breed animals show qualities of both purebred parents.

HOW BEEF CATTLE ARE RAISED

Farmers raise beef cattle on large pastures, or **rangeland** (RAYNJ land). They can also use small, fenced areas.

Beef cattle spend their first months on rangeland. Then, farmers round them up and transport, or drive, them to **feedlots** (FEED lots). In feedlots, farmers feed cattle a mix of grains and hay for four to eight months.

The young cattle may gain 500 to 600 pounds (227 to 273 kg) in the feedlots. Because they are fed and fattened in the feedlots, these young animals are called "feeder cattle."

Beef cattle graze on rangeland in central Florida.

A butcher uses a band saw to cut through beef bone. Butchers also use knives to process beef carcasses.

Brahman cattle, like this purebred Brahman bull, and mixed breeds with Brahman blood do well in hot climates.

READY FOR MARKET

Beef cattle are usually **slaughtered** (SLAW terd), or killed, when they are 13 to 16 months old. They weigh about 1,200 pounds (545 kg).

Just before slaughter, beef cattle are called "finished" cattle. The farmer has finished raising them. They are the ideal age and weight to be **processed** (PRAH sest), or changed, into meat.

Farmers gather finished cattle from the pens and load them onto trucks. The trucks haul them to a slaughterhouse or meat-processing plant.

A truck delivers food to cattle troughs at a Midwestern feedlot. These mixed-breed cattle are being finished for market.

CATTLE TO BEEF

At the plant, cattle are slaughtered. Their skin and insides are taken away. What's left, the meaty part, is a **carcass** (KAR kass). The carcasses are hung in cold storage rooms. A 1,200-pound (545-kg) cow has about 450 pounds (205 kg) of meat on its carcass.

State meat inspectors watch the slaughter and cleaning. Inspectors make sure that meat is healthy and the plant is clean. Inspectors also make sure that cattle are killed instantly.

A meat-processing plant owner checks over fresh beef carcasses in a cold storage room. Each carcass bears a stamp from a state beef inspector.

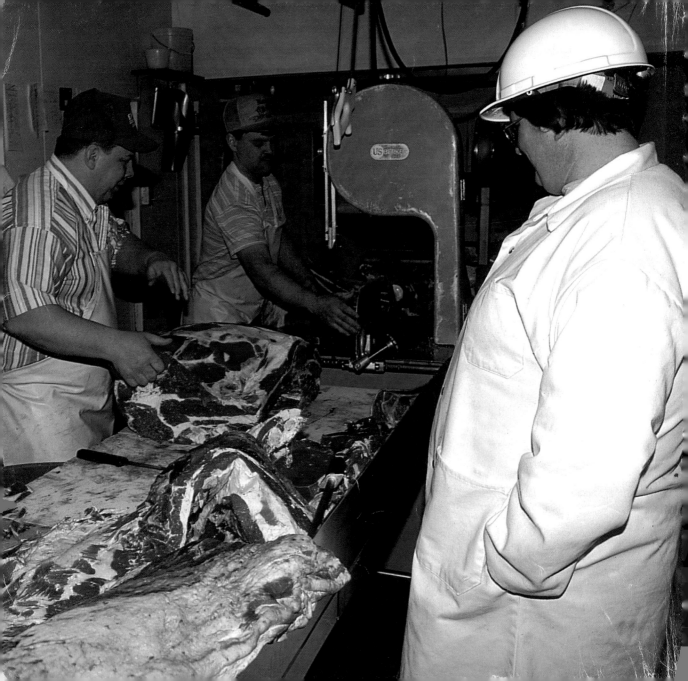

PROCESSING

A carcass hangs in cold storage for about a week. Butchers then cut it into pieces. Large pieces, such as a half carcass, are sent to supermarkets and meat markets. There the meat will be cut into smaller pieces—steaks and roasts, for example.

Butchers cut some meat into smaller pieces at the processing plant. It may be shipped to restaurants or stores.

BEEF PRODUCTS

The best known use of beef in North America is hamburgers. Hamburger is ground beef. Hamburger meat also has fat, which sizzles in the fry pan.

Some beef is used in "processed meats," like hot dogs and bologna.

Nothing is wasted when cattle are slaughtered. Parts of the animal are used for things such as animal foods and fertilizers. The skin is used for leather.

A butcher at a meat-processing plant grinds hamburger. Hamburger has more fat than other cuts of beef.

BEEF AS FOOD

Beef is a food rich in **nutrients** (NU tree ents). Nutrients are what our bodies need for energy, strength, and growth.

Beef has large amounts of protein. Protein is an important food for the human body. In addition, beef has many minerals and B vitamins.

Beef also has fat and **cholesterol** (koh LES te rall). Cholesterol is a very fatty substance. *Too much* fat and cholesterol in someone's diet can lead to heart and blood problems.

Glossary

breed (BREED) — a particular group or type of farm animal within a larger group of very closely related animals, such as *Hereford* cattle among all cattle

carcass (KAR cass) — the body of an animal being processed into meat

cholesterol (koh LES te rall) — a waxy, fatty substance in human blood

feedlot (FEED lot) — a fenced area where cattle are fed before slaughter

nutrient (NU tree ent) — any of several "good" substances that the body needs for health, growth, and energy; vitamins and minerals

processed (PRAH sest) — changed in form or moved from one place to another; prepared for market

rangeland (RAYNJ land) — a huge area of open land livestock use to graze

slaughtered (SLAW terd) — to have been killed for food

INDEX